Senses

Ready, Set... STEM!

By William Anthony

©This edition published 2024.
First published in 2021.
BookLife Publishing Ltd.
King's Lynn, Norfolk
PE30 4LS, UK

HB ISBN: 978-1-83927-838-9
PB ISBN: 978-1-80505-677-5

Written by:
William Anthony

Edited by:
Robin Twiddy

Designed by:
Amy Li

All rights reserved. Printed in India.

A catalogue record for this book is available from the British Library.

All facts, statistics, web addresses and URLs in this book were verified as valid and accurate at time of writing. No responsibility for any changes to external websites or references can be accepted by either the author or publisher.

Photo Credits

Images are courtesy of Shutterstock.com. With thanks to Getty Images, Thinkstock Photo and iStockphoto.

Recurring images – Micra (brain, sunburst pattern), Lorelyn Medina (children vectors), illustrator096 (clouds). Cover–p1 – KittyVector, Beskva Ekaterina, Inspiring, Oxy_gen, svtdesign, p2–3 – Samuel Borges Photography, Kitty Vector, Oxy_gen, p4–5 – Africa Studio, svtdesign, LuckyVector, Toey Toey, Inspiring, p6–7 – govindamadhava108, Mila Ma, Pogorelova Olga, Studio Barcelona, p8–9 – Panda Vector, ilikestudio, Studio_G, Gaidamashchuk, p10–11 – Faberr Ink, Anatolir, svtdesign, Perfect Vectors, p12–13 – Infinity, Olga Shlyakhtina, Seika Chujo, p14–15 – Inspiring, Pro Symbols, Kolpakova Daria, Alexander Lysenko, p16–17 – Ksusha Dusmikeeva, Rido, Alexander Lysenko, p18–19 – Rudmer Zwerver, TY Lim, Sko Helen, p20–21 – 32 pixels, Alexander Lysenko, Lukiyanova Natalia frenta, p22–23 – goodluz, ANURAK PONGPATIMET, p24 – Inspiring.

Contents

Page 4 The World around Us
Page 6 Sniff, Sniff
Page 8 Ear to Hear
Page 10 Eye, Eye, Captain
Page 12 Yum, Yum
Page 14 Touchy Feely
Page 16 Senses at School
Page 18 Be Like a Bat
Page 20 Superhero Senses
Page 22 A World of Senses
Page 24 Glossary and Index

Words that look like <u>this</u> can be found in the glossary on page 24.

The World around Us

Hi! I'm Brain the brian. No, wait. I'm Brian the brain — that's it. That name has always given me trouble! Let's learn about senses together!

We use senses to understand the world around us. We have five main senses. They are smelling, hearing, seeing, tasting and touching.

We use different parts of our body for each sense.

Sniff, Sniff

We use our noses to sniff and smell things. Some things smell nice, but other things might stink!

What is your favourite smell?

Let's find something to smell. Take a walk outside. Can you find a flower or a pretty leaf? Hold it to your nose and…

Ready, Set… SMELL!

What did it smell like? Was it nice or not?

Ear to Hear

We use our ears to hear different sounds around us. The sounds go into our ears and bounce off our <u>eardrums</u>.

Your eardrum is inside your ear, but never try to touch it.

Let's test out different sounds. Grab a pencil and some different objects. Tap your pencil on each one to see how each one sounds. On your marks…

Ready, Set… TAP!

Ear

Eye, Eye, Captain

When you look around, you are using your eyes to see. You see an object because light has bounced off it and into your eyes.

Light

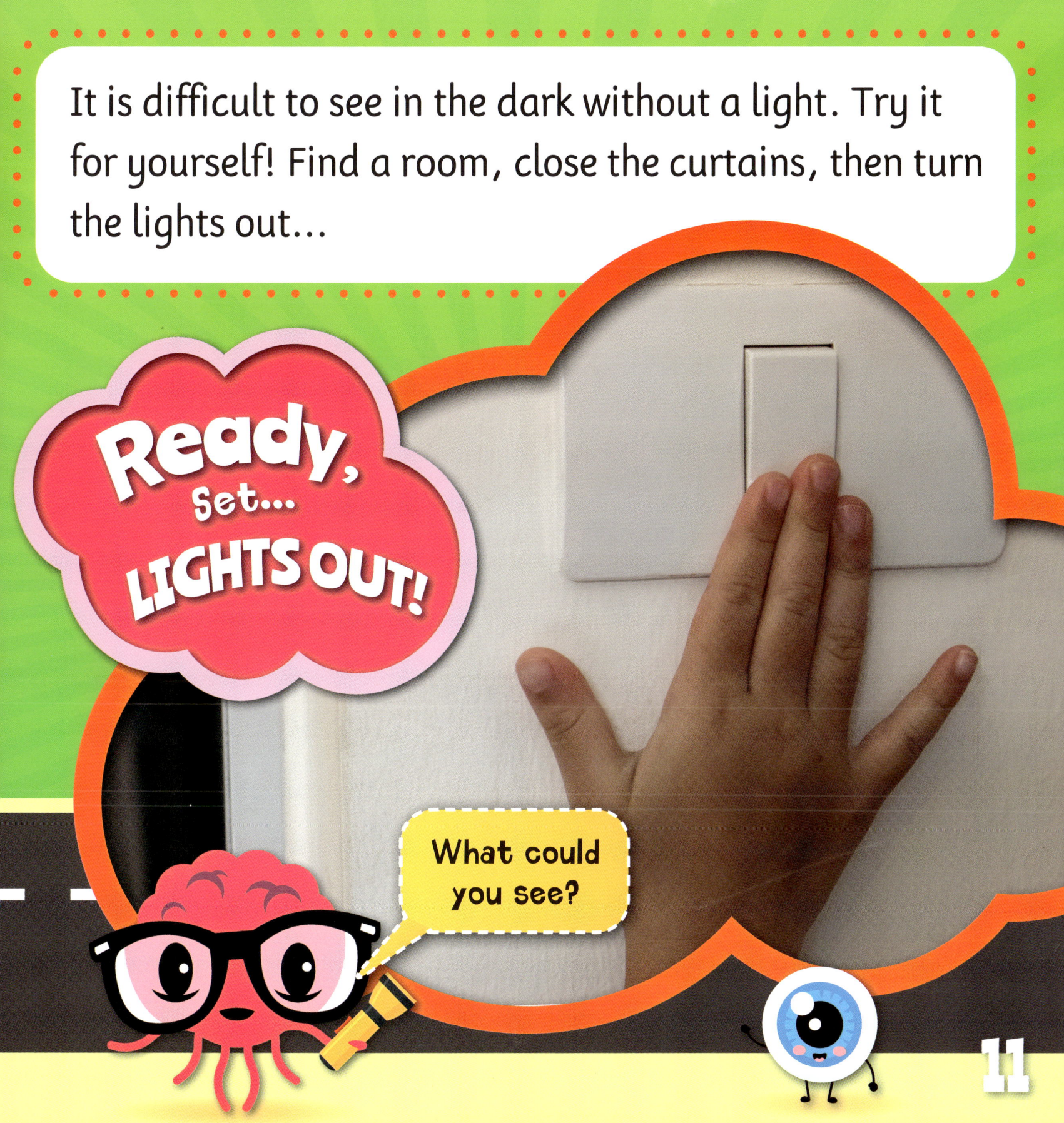

Yum, Yum

You can taste things using your tongue. Your tongue has lots of taste buds on it. These help you to know whether something might be sweet, sour or something else.

What is your favourite flavour? Mine is raspberry!

Let's do a taste test! Grab two pieces of fruit, such as a strawberry and a lemon. Taste both and try to say how they are different. Have you got your fruit?

Ready, Set... TASTE!

Touchy Feely

Touch is how we feel things. Our skin is very <u>sensitive</u> and can tell us how something feels.

We can feel things with lots of different parts of the body.

Let's test out our touch. Look around the room. Try to find:

- Something soft
- Something hard
- Something rough
- Something smooth
- Something warm
- Something cold

Ready, Set... TOUCH!

Senses at School

We use our senses all the time. We use them at home, at the park, in school and everywhere we go.

What do you use your senses for at school? Make a list of everything you might use your senses for.

Use this example to get you started.

Ready, Set... LIST!

Smell	• New books
Hear	• Teacher
See	• Table
Taste	• Lunch
Touch	• Pencil

Be Like a Bat

Some animals have amazing senses. Did you know that bats have incredible hearing? They use their hearing to find out where things are in the dark.

Let's try being like a bat!

Find an adult to help you. Close your eyes so you cannot see and let your adult guide you into a room. Try to use your senses such as touching, hearing and smelling to work out where you are.

Ready, Set... EXPLORE!

Superhero Senses

Have you ever wanted to be a superhero? Lots of superheroes have super senses. Draw yourself as a superhero with a super sense.

What is your super sense? Can you write a sentence about it?

Which super sense did you choose? Can you see through walls or see in the dark? Can you hear things from very far away? What about smelling danger?

A World of Senses

Some people cannot use all their senses. People who are blind find it difficult to see. They might have a guide dog to help them get around.

People who are deaf find it difficult to hear. Some deaf people might use a hearing aid to help them hear the world around them.

Many people who are <u>disabled</u> use all sorts of things to help them sense the world.

Glossary

disabled	finding certain things difficult because of a medical condition
eardrums	the parts of ears that vibrate when sounds hit them
sensitive	quick to work out how something feels
sour	a strong, sharp taste
taste buds	the small spots on your tongue that let you taste things

Index

animals 18
disabilities 22–23
hearing 5, 8, 17–19, 21, 23
school 16–17
seeing 5, 9–11, 17, 19, 21–22
smelling 5–7, 17, 19, 21
tasting 5, 12–13
touching 5, 8, 14–15, 17, 19